WITH A POCKET CALCULATOR

by Rose Wyler and Mary Elting
Pictures by G. Brian Karas

JULIAN MESSNER

Acknowledgments

For good advice and practical help the authors thank
Dr. Roberta Flexer
Dr. Robert Moll, Associate Professor of Computer Science, University of Massachusetts, Amherst, Massachusetts
Audrey Tebrich who studies math at Southern Hills Junior High School, Boulder, Colorado.

Published by Julian Messner,
a division of Simon & Schuster
Simon & Schuster Building, Rockefeller Center
1230 Avenue of the Americas
New York, New York 10020

JULLIAN MESSNER and colophon are trademarks
of Simon & Schuster

10 9 8 7 6 5 4 3 2 1 (hardcover)

10 9 8 7 6 5 4 3 2 1 (paperback)

Library of Congress Cataloging-in-Publication Data

Wyler, Rose.
Math fun with a pocket calculator/by Rose Wyler and Mary Elting.
p. cm.
Summary: Instructions for playing a variety of games and tricks with a pocket calculator.
1. Mathematical recreations—Juvenile literature. 2. Calculators—Juvenile literature. [1. Mathematical recreations. 2. Calculators.] I. Elting, Mary. II. Title.
QA95.W948 1992
793.7'4—dc20
ISBN 0-671-74308-2 (library) ISBN 0-671-74309-0 (paper)
91-16265
CIP
AC

Hi!

Do you know that your calculator is a tool in a class all by itself? You'll use it for work—but you'll also discover that it is a marvellous invention for having fun. It's as if a mathematical wizard has given you a special way to play games, do tricks, and even make your remarkable little machine "talk."

Maybe you have wondered why the numbers in your calculator display have strange shapes. The Math Wizard can tell you that there's a good reason for those shapes. Each number can be formed by straight lines. And each line can be made to appear by turning on a tiny light bulb behind the calculator display window. There are usually 8 or 9 places in the display where a number flashes on. Behind each place is a double square made up of seven bulbs. With only those seven bulbs, any digit from 0 through 9 can be formed. Special tiny pathways connect the bulbs to the calculator buttons, the way electric wires connect light switches to lamps in your house. Push a button and the lights go on.

Your calculator works problems in its own special way. Suppose you give it this to do: $5 + 6 \times 8 = ?$ Your display may show the answer 88. But your friend's display may show 53! What is the difference?

Your calculator works problems in the order in which you punch the buttons. *Your friend's* does the multiplication or division first, and then the addition or subtraction. But *both calculators* will give the same answer if both of you press the equals button after each operation.

The games and tricks in this book are arranged so that all calculators will give the same answer if the equals button is pressed between operations. That is: $5 + 6 = 11 \times 8 = 88$.

Rose Wyler and Mary Elting

TABLE OF CONTENTS

MONKEY BUSINESS

If a monkey played around long enough, poking typewriter keys, could it eventually write *Tarzan of the Apes?* Maybe not. But by guessing and testing, you *can* make your calculator solve some monkey-business puzzles. Get ready to meet the Math Wizard. He's here to help you have more fun than a barrel of monkeys with the following puzzles.

FICKLE FIVES

"I'm thinking of the number 5," said the Math Wizard. "It can't seem to make up its mind. Sometimes three 5s get together and arrange themselves with a plus or minus or multiply or divide sign, so that they result in 0. But sometimes they equal 2. And other times they just end up being 5."

Experiment with your calculator until you get the signs in the right places.

5 5 5 = 0

5 5 5 = 2

5 5 5 = 5

Is this what you get?

$5 - 5 \div 5 = 0$

$5 + 5 \div 5 = 2$

$5 \div 5 \times 5 = 5$

also $5 + 5 - 5 = 5$

TALE OF A TAIL

The Wizard has a monkey. One day, he and his monkey were planting flowers in the back yard when a cat came to see what was going on. Unfortunately, the cat had fleas. A flea jumped from the cat and landed at the tip of the monkey's tail. But the flea didn't bite. Instead, it decided to climb the monkey's tail.

The monkey was sitting still when the flea started its climb. But every once in a while it would give the tail a little flip, and that made the flea slide downward. Each time the flea slipped, it tried again. Finally, it reached the monkey's back.

The Wizard had been watching the flea, and he made up this puzzle about it:

The monkey's tail is 16 inches long. With every jump the flea climbed 4 inches up the tail. But every time the monkey flipped its tail, the flea slipped down 2 inches. How many jumps did the flea have to make in order to reach the monkey's back?

Here's how to find out: To calculate the answer, take it one climb and flip at a time. The first climb (4) minus the first slip back (2) leaves it 2 inches ahead. That 2 inches, added to the next climb (4) puts it 6 inches ahead, but then it slips back 2 and is only 4 inches ahead. Here are the calculations:

$0 + 4 = 4$	$4 + 4 = 8$	$12 - 2 = 10$
$4 - 2 = 2$	$8 - 2 = 6$	$10 + 4 = 14$
$2 + 4 = 6$	$6 + 4 = 10$	$14 - 2 = 12$
$6 - 2 = 4$	$10 - 2 = 8$	$12 + 4 = 16$
	$8 + 4 = 12$	

Now count up the number of times the flea jumped forward (each time you added 4). The answer is seven jumps.

NINJA FLAMINGOS

One day the Wizard took his monkey to the zoo. After they visited the gorillas and chimps, they went to see the turtles and flamingos. A lot of people were watching the animals, and it was hard to see them all. But the Wizard and the monkey counted 24 animals and a total of 78 animal legs.

How many flamingos and how many turtles were in the zoo?

Hint: Start experimenting with a number for the turtles and guess and test on from there. Remember that turtles have four legs; flamingos have only two.

Answer: The monkey finally figured out that he saw 9 flamingos and 15 turtles. However, the zookeeper said that the Wizard had made a mistake. He counted wrong because people got in the way. There were, indeed, 15 turtles with four legs each, making 60 turtle legs. But there were 10, *not* 9 flamingos. Two of them were standing on only one leg.

MYSTERY NUMBERS

The Wizard said to his monkey, "Arrange eight 8s so that the result is 1,000."

It took his monkey four days to solve the puzzle.

How fast can you do it?

"Now," said the Wizard, "arrange seven 4s so that the result is 100."

The monkey did it in two days. How about you?

Hint: Both these puzzles can be solved by addition of whole numbers.

Answers:

888 + 88 + 8 + 8 + 8 = 1,000
44 + 44 + 4 + 4 + 4 = 100

And now try your skill on this one. It's a little harder. Each question mark between numbers stands for either a plus sign or a multiplication sign.

9 ? 8 ? 7 ? 6 ? 5 ? 4 ? 3 ? 2 ? 1 = 100

Hint: Aim for a pretty big number early in your calculation.

Answer:

$9 \times 8 + 7 + 6 + 5 + 4 + 3 + 2 + 1 = 100$

SEEING DOUBLE

One day, Ellie was visiting the Wizard and his monkey. Suddenly, the Math Witch walked in with *her* monkey.

"My monkey is getting smarter and smarter," said the Wizard. "I gave him this addition problem, and he did it in no time. He replaced the symbols with numbers and got the right answer."

```
  □ ◇ 9
+ ♧ 1 ♡
-------
  5 6 7
```

"My monkey is just as smart as yours," snapped the Witch. "She can do the problem in less than no time."

After a while the Wizard showed Ellie what his monkey's answer was. Then the witch showed *her* monkey's answer.

"They're different!" Ellie cried. *"And they're both right."*

Explore with your calculator and find out what Ellie saw.

Answers:

The Wizard's monkey added

```
  249
+ 318
-----
  567
```

The Math Witch's monkey added

```
  349
+ 218
-----
  567
```

The equal sign we use (=) is only about 400 years old. The man who invented it, Robert Recorde, may have been thinking of parallel lines, which are always an equal distance apart. Most teachers and mathematicians began using Recorde's sign right away. But a few stubborn people kept on writing equal signs like this:

MAMA MIA!

What's special about a mom? Sure, she's a nice lady, but the word *mom* is also a palindrome—a word that reads the same backward and forward. There are palindrome numbers, too—121, for example. And there are secrets to making them. Try this with your calculator:

1. **Enter a three-digit number in which the first digit on the left is greater than the last digit on the right, for example, 534.**
2. **Reverse the number and add it to the first number: 534 + 435 = 969.**
3. **Perform steps 1 and 2 with another such three-digit number, for example, 341 + 143 = 484.**

Now comes the interesting part of this monkey business. You may not always get a palindrome on the first addition. But if you keep on reversing and adding the numbers you get, you will eventually end up with a palindrome!

Try this: Start with a two-digit number, such as 68 or 97. Reverse and add, reverse and add three or more times and see what you get.

Answers:

68	97
+ 86	+ 79
154	176
+ 451	+ 671
605	847
+ 506	+ 748
1,111	1,595
	+ 5,951
	7,546
	+ 6,457
	14,003
	+ 30,041
	44,044

With four digits you can make calendar-year palindromes. The only one in the twentieth century was 1991. How many will there be in the twenty-first century?

You guessed it! One—2002.

People in Europe invented plus and minus signs long ago. Bookkeepers borrowed "+" from writers who got tired of spelling out the Latin word for *and*. That is why we still say "three and three make six." About 500 years ago, bookkeepers also got tired of spelling out *minus*. They shortened it to "ms," then to "m̄," and finally just dropped the "m." That's why we write "6 − 3."

OKAY, THIRTY-SEVEN!

Do you have a least favorite number? Carmen did.

"I don't like the number 37," Carmen told the Wizard. "It's boring. You can't do anything with it. It's not special like 36, which can be divided by a lot of other numbers."

"*Hmm,*" said the Wizard. "I think you'll change your mind if you do these two puzzles."

1. **Write 37 using five 3s, with two plus signs and one division sign.**
2. **Write 37 using five 3s, with one multiplication and one division sign.**

Answers:

$3 \div 3 + 33 + 3 = 37$

$333 \div (3 \times 3) = 37$

Carmen still wasn't convinced.

"All right," said the Wizard. "Try this."

1. **Think of a three-digit number in which all the digits are the same—for example, 222.**
2. **Add the digits: $2 + 2 + 2 = 6$.**
3. **Divide the number in step 1 by the number in step 2—that is, $222 \div 6 = 37$.**
4. **Now perform those same steps with all the other such three-digit numbers.**

Did Carmen change her mind? Do the calculations, and see what you think.

Answers: Carmen had to admit that 37 is a special number!

$111 \div 3 = 37$	$666 \div 18 = 37$
$222 \div 6 = 37$	$777 \div 21 = 37$
$333 \div 9 = 37$	$888 \div 24 = 37$
$444 \div 12 = 37$	$999 \div 27 = 37$
$555 \div 15 = 37$	

Quizzle: Is this a picture of a domino? Find out on page 14.

DOMINO DEAL

"Did you know that I have suddenly acquired magic power over your calculator?" José said to Ginny.

"Show me," Ginny answered.

José told her to take a domino out of the box. "Don't show me what it is. Now do the calculations I'm going to dictate."

1. **Enter the larger of the two numbers on the domino. Suppose it is 4.**
2. **Multiply it by 2: $4 \times 2 = 8$.**
3. **Multiply by 5: $8 \times 5 = 40$.**
4. **Add 6: $40 + 6 = 46$.**
5. **Multiply by 2: $46 \times 2 = 92$.**
6. **Subtract 12: $92 - 12 = 80$.**
7. **Divide by 2: $80 \div 2 = 40$.**
8. **Add the smaller number on the domino. Suppose it is 1: $40 + 1 = 41$.**

"That's all," José says. He takes the calculator from Ginny, looks at the display and tells her, "The dots on the domino are 4 and 1."

He is right! José knows the answer by reading the display: the larger number in the display is 4 and the smaller is 1.

The Secret: Steps 2 and 3 put the larger number of dots in the tens place. Steps 4, 5, 6, and 7 just put numbers in and took them slyly out. Step 8 puts the smaller number in the ones place.

What if Ginny tried to play a trick on José and picked a domino that was a double? That way neither end would be the larger or smaller. José didn't have to worry. If the domino was a double, for instance 6, the display would show 66, and he would be able to "mysteriously" know that neither number was larger.

WHAT'S THAT AGAIN?

"Does your calculator ever stutter?" you ask Katie.

Katie probably thinks that's a silly question.

"I bet you a penny I can make it stutter," you say. Then give Katie this problem:

1. Enter 15,873.

2. Multiply by 21.

"Now read the display out loud," you tell Katie.

"Three-three-three-three-three-three."

You win a penny.

HERE'S ANOTHER ONE

"I can make your calculator think it's a railroad train," you tell Tony. "Just have it do this problem:"

1. Enter 1 2 3 4 5 6 7 . 9. (Notice that you use a decimal point instead of 8 between 7 and 9.)

2. Multiply by 18.

"Now read out loud what's in the display."

"Two-two-two-two-two-two-two-two," says Tony.

"What did I tell you? *Toot-toot-toot!*"

FIDDLING AROUND

Here are some more brain teasers:

1. The Wizard said, "I'm thinking of an even two-digit number. It is divisible by an odd number, and the sum of its digits is divisible by 3. Now what is the even two-digit number and what is the odd number?"

Hint: Start with the question: What even numbers are divisible by 3? (Answer on page 14.)

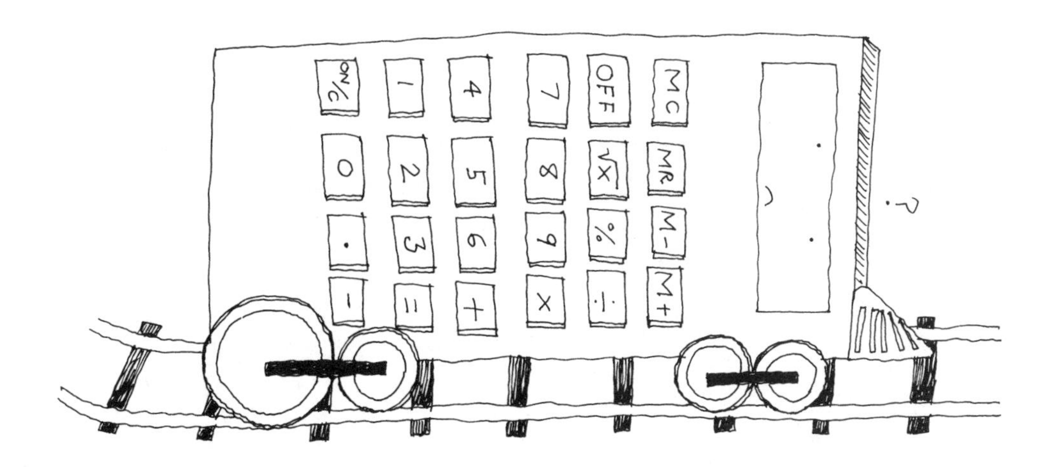

Answer: There are a lot of answers to this. Some of them are:

18 and 9
36 and 9
54 and 9
72 and 9
Can you go on from there?

2. The Witch said, "I'm thinking of three consecutive numbers that add up to 48. What are they?"

Estimate and experiment with your calculator.

Answer: 15, 16, 17

3. The Wizard said, "I'm thinking of four consecutive numbers that add up to 98. What are they?"

Estimate and experiment.

Answer: 23, 24, 25, 26

4. Here is that strange number again: 1,234,567.9

Try these experiments with it.

Multiply it by 9. Multiply it by 27. And look what happens when you multiply it by 8, which is the missing number in the sequence.

You can discover a lot of patterns by experimenting with numbers.

Is this a sketch for a new kind of display in a calculator? Guess again. About 700 years ago, arithmetic teachers used it to help students remember how to write numbers. Before that, people wrote the first nine digits in different ways, and that often caused confusion. So everyone began learning the new style, which is much like the one we use today.

Quizzle Answer: Did you see a domino in the picture on page 12? The Wizard says it is two kittens in a tunnel.

GO TEAM!

Northside School (black uniforms) is playing Eastside School (white uniforms) for the football championship. Your problem is to draw a straight line dividing the field so that the same number of players is on each side of the line—*and* the numbers on the players' backs on each side of the line must add up to the same two-digit number.

Your calculator will tell you if you have the line in the right place.

Turn the page to find the dividing line and the two-digit number.

Does your calculator tell you 66 is the secret number?

DIGGING DIGITS

"I'm thinking of two two-digit numbers," the Wizard says. "They have the same digits and one is the reverse of the other—21 and 12, for example. When you add together the digits in each of my numbers, you get 10. The difference between my two numbers is 18. What are my two numbers?"

Try to estimate the answer first. Then guess and test.

Hint: Start with the biggest two digits that make 10 and work down from there: 9 + 1 = 10; 91 − 19 = 72; 82 − 28 = 54; 73 − 37 = 36; and so on.

Answer: 64 − 46 = 18

PIZZA HUNT

One day, the Witch decided to give a pizza party for all the kids in the neighborhood. Before the party started, there would be a pizza hunt. The first person to choose the right route to the park would win a prize.

Here is how the hunt would work: The Witch would go to every house in the neighborhood and put a number on each door. Some numbers would have a plus sign and some would have a minus sign. The hunt would start at the house on the northwest corner of Easy Street. From there each child would choose a route from one house to another by knocking at certain doors, until he or she ended up in Cheeseman Park. Using a calculator, each

child would add or subtract the number on the door where he or she knocked. In order to win, a child would have to finish with exactly 100 in the calculator display.

Study the map of the neighborhood, then explore with your calculator until you add and subtract just the right numbers to create 100. Can you win the pizza hunt?

What was the Witch's prize? An extra big pizza, of course. She also gave a booby prize to the last person to reach the park—a box of chocolate-covered grasshoppers.

The winner was Sid who had a pizza hound named Pepperoni. The hound had such a good nose that he sniffed out the route right away. Look for it on page 18.

PATCHWORK HOMEWORK

Maria decided to do her math homework while she was minding her little brother. She gave him some crayons to play with, and then the phone rang. It was Jerry. She and Jerry talked for *quite* a while. All that time little brother was amazingly quiet. No wonder. When Maria came back, she found that he had been decorating the problems she had already checked with her calculator. Three of them looked like this:

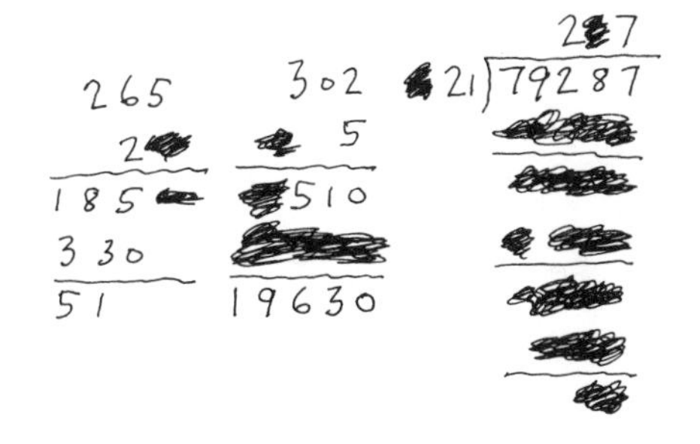

By guessing and testing, Maria fixed the problems up. Can you? See the answers on page 18.

About 800 years ago an Arab mathematician said, "Let's write 4 divided by 2 like this: 4/2." However, some people insisted on making the sign this way "·|·". Others preferred "÷" or just two dots like this ":". Mathematicians still don't agree. In England, the United States, and some other English-speaking countries "÷" means "divided by." In Europe and Latin America the two-dot sign ":" is almost always used.

Answers:

```
  265      302          247
 × 27     × 65   321)79,287
1,855    1,510        642
  330    1,812      1,508
5,155   19,630      1,284
                    2,247
                    2,247
                        0
```

HOW SID WON THE PIZZA HUNT

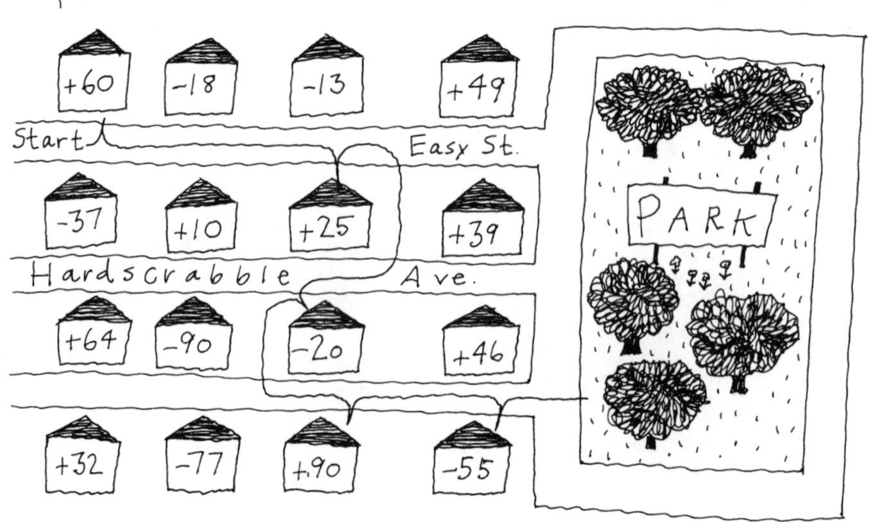

BUTTON MAGIC

The buttons on your calculator are arranged in a special pattern. You can experiment with the numbers on the buttons and make an amazing discovery.

1. **Enter the numbers in the bottom row from right to left: 321**
2. **Subtract the numbers in that row from left to right: 321 – 123**
3. **On paper write the number in the display. Did you get 198?**

Press "clear" and do this:

1. **Enter the numbers in the middle row from right to left: 654**
2. **Reverse the numbers in the middle row and subtract: 654 – 456**
3. **Write down the number that appears in the display.**

Surprise—it's 198 again! Press "clear" and do this:

1. **Enter the numbers in the top row from right to left: 987**
2. **Reverse and subtract: 987 – 789**
3. **What is the number in the display this time?**

Curiouser and curiouser. Try some other subtractions. What about making crosses such as: 951 and 753; 852 and 654?

BIG STUFF

If it's big, it's special—Brontosaurus, a million dollars, a billion stars. Get ready for some special big numbers—they're part of some really big experiments with your calculator.

CAR WASH CAPER

Suppose you tell Mr. Pinchpenny that you will wash his car once every two weeks if he will pay you 1 cent for the first wash, 2 cents the next time, 4 cents the next time and so on, doubling the number of pennies with each wash. He will probably think that's a great bargain.

How good a bargain do *you* think it is? Let's see how much you will be making when you finish the tenth wash:

$1 \times 2 \times 2 \times 2 \times 2 \times 2 \times 2 \times 2 \times 2 \times 2 = 512$ pennies.

Not spectacular, but not too bad.

Now keep on calculating, multiplying each time by 2. Your fifteenth wash will be worth 16,384 pennies. Now you're getting somewhere.

Your twentieth will bring 524,288 pennies.

Your twenty-fifth will bring 16,777,216 pennies.

How much will Mr. Pinchpenny have to pay you for the twenty-seventh wash? Estimate, then use your calculator.

Answer: 67,108,864 pennies, or $671,088.64!

OVERLOAD!

Suppose you want to know how much you will earn if you can persuade Mr. Pinchpenny to pay you for still another car wash. Try to multiply $671,088.64 by 2, and what do you get? Probably a signal that there is no room for a nine-digit number in the display. Most calculators have only eight places, plus a place for the decimal point. But don't give up. There is a way to get around the difficulty. It is called overload. Using paper and pencil, you help your calculator a little.

1. **Separate this big number into two parts, the dollars and the cents: 671,088 and .64.**
2. **Using your calculator, multiply each part by 2: 2 × 671,088 and 2 × .64.**
3. **Record the products on a piece of paper. Be sure when you write them down that the decimal points are aligned, so that the place values correspond:**

millions	hundred thousands	ten thousands	thousands	hundreds	tens	ones		tenths	hundredths
1	3	4	2	1	7	6	.		
						1	.	2	8
1	3	4	2	1	7	7	.	2	8

A SHOE-IN

There were no calculators in the year 1535 when a similar problem appeared in an arithmetic book. But the story was different. People in those days rode on horseback or in horse-drawn carriages and wagons, and their horses had metal shoes on their hooves. When they had to get new shoes, the blacksmith fastened them on with nails. Each shoe was put on with 6 nails.

It happened that a certain blacksmith had studied arithmetic, and when the king brought in his favorite riding horse to be shod, the blacksmith offered him this deal: He would charge 1 coin for the first nail, 2 coins for the second, 4 coins for the third, 8 coins for the fourth and so on till the job was done. The king was a miser, and he never read math books. So he agreed. Was he surprised when the blacksmith gave him the bill!

How many coins did the king have to pay?

Hint: A horse has 4 feet. Each foot received a shoe that needed 6 nails. This means that the blacksmith used altogether 6 × 4 = 24 nails. The first nail cost the king 1 coin; the second 2 coins.

Keep on doubling the number of coins until all 24 nails are in place. Estimate and then do the calculating. Look for the answer on page 22.

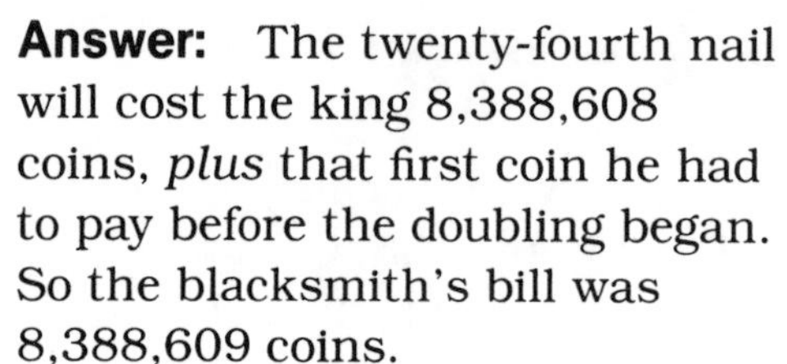

Answer: The twenty-fourth nail will cost the king 8,388,608 coins, *plus* that first coin he had to pay before the doubling began. So the blacksmith's bill was 8,388,609 coins.

ALL THE KING'S HORSES

Suppose that the same king also decided that he might as well order new shoes for the queen's horse and for the 12 horses that pulled his royal carriage. The blacksmith said he would only charge as much for each of these 13 horses as he charged for the riding horse. Of course, he did not tell the king that his total bill would be: $8,388,609 \times 13 = 109,051,917$ coins *plus* the bill for the king's own riding horse which was already 8,388,609 coins.

How can you add those two big numbers? If you try to do it with your calculator, you will get an overload signal. Don't despair. You can easily do big addition with the help of paper and pencil, working the problem in three steps:

1. **Separate each number between the hundreds and the thousands: 109,051 and 917; 8,388 and 609.**
2. **Using your calculator, add the corresponding digits:**

109,051	**917**
+ 8,388	**+ 609**
117,439	**1526**

3. **Record the sums on a piece of paper. Be sure to align your digits in their proper place value, then add.**

hundred millions	ten millions	millions	hundred thousands	ten thousands	thousands	hundreds	tens	ones
1	1	7	4	3	9			
					1	5	2	6
1	1	7	4	4	0	5	2	6

The blacksmith's bill was 117,440,526 coins!

Quizzle: Suppose there is an orchard of 172 apple trees, each averaging 332 branches, each branch with 41 leaves at the beginning of May. New leaves are added to each branch at the rate of 10 a week. How many leaves will be on the tree after 8 months?

Guess, then look on page 24.

NUMBERS IN YOUR LIFE

Your heart is only about the size of your fist, but it does a big job. How does it work? The heart is really a kind of pump made of muscle. Its job is to push blood through about 60,000 miles of tubes called arteries, veins, and capillaries in your body.

Your heart muscle squeezes together about 80 times a minute to give the blood a push. Then the muscle relaxes. That is what is called a heartbeat. Day and night, for as long as you live, your heart goes on beating, about 80 times a minute. Your calculator can tell you how many times it beats at that rate in a year. Be prepared for a big number.

Hint: There are 60 minutes in an hour; 24 hours in a day; 365 days in a year.

Now turn the page.

Answers: In an hour your heart beats 60 × 80 = 4,800 times; in a day it beats 4,800 × 24 = 115,200 times; in a year it beats 115,200 × 365 = 42,048,000 times.

Quizzle answer: None. Trees shed their leaves in the fall.

When people first began doing arithmetic in Europe, there was no one multiplication sign. One writer tried to promote the idea of a star-shaped symbol something like our asterisk (*). It wasn't very popular then, but today computer people use "*" to indicate multiply. About 300 years ago, an Englishman named William Oughtred began using a cross, and it caught on. Later, especially in England, writers of math books complained that the cross looked too much like the letter "x" which they were using as a sign for "unknown number." So they used a dot between numbers: 2 . 2. In the United States, when the dot is used, it is raised a little: 2 • 2.

MORE NUMBERS IN YOUR LIFE

Your heart beats all by itself. You never have to think about it. How does it work? It actually has its own little electric motor! The motor, called a pacemaker, is a small group of special cells that create tiny surges of electricity. The electric current makes the heart muscle contract. If you live to be eighty years old, your wonderful heart will beat more than 3,000,000,000 times.

STILL MORE NUMBERS

There is about a gallon of blood in a grown-up's body. The heart pumps it out and back many times a day. If a water pump in a well did as much work as the heart, it would pump out about 2,000 gallons in a day.

1. **If your heart pumps 2,000 gallons in a day, how much does it pump in a year?**
2. **If a person lives to be 70 years old, about how many gallons will his or her heart pump in a lifetime?**

Answers:

1. There are 365 days in a year, so in a year the heart pumps $365 \times 2{,}000 = 730{,}000$ gallons.
2. There are about 17 leap years in 70 years, so we have to do this answer in two parts:

$70 \times 730{,}000 = 51{,}100{,}000$ gallons in 70 ordinary years.
Now we calculate the number of gallons in 17 leap-year days:
$17 \times 2{,}000 = 34{,}000$.
Finally we add:
$51{,}100{,}000 + 34{,}000 =$
51,134,000 gallons in a lifetime.

POTHOLES, POTHOLES

The school bus lurched and Julio bounced halfway out of his seat. The driver had not seen the pothole in the road. The front wheel was damaged, and Julio was late for school. His teacher was not surprised. She knew all about potholes, and she gave the class this problem.

"Every year," she said, "there are 1,500,000 new potholes in the streets of New York City. It takes about 20 minutes to clean out the cavity in the road and fill it with new asphalt. Now take out your calculators and discover the answers to these questions:"

1. **How many minutes does it take to fix all of the city's potholes?**
2. **How many hours does it take to fix them?**
3. **If a man worked 8 hours a day, how many days would it take for him to do the job alone?**

Answers:

1. **If each pothole takes 20 minutes to fix, it will take $20 \times 1{,}500{,}000 = 30{,}000{,}000$ minutes to fix all of them.**
2. **There are 60 minutes in an hour; so it will take $30{,}000{,}000 \div 60 = 500{,}000$ hours to fix them.**
3. **One man, working 8 hours a day, would spend $500{,}000 \div 8 = 62{,}500$ days doing the job.**

If he worked 300 days every year, he would be more than 200 years old before he finished! That's why New York has more than 500 workers with 130 machines at work fixing potholes.

TEASERS, TALES, AND TRICKS

Some of the tales you are about to read are tricky, so watch out.

CALCULATOR GREETING CARD

Here is a way to wow your mom on her birthday. Suppose her birthday is June 21. First thing in the morning, hand her your calculator and say, "Here—it has a message for you. Just turn it on. Now enter 50. Subtract the number of days in this month. Add 1. That's it."

You look at the number in the display. "Twenty-one," you say. "It's telling you something special. What's today? The twenty-first?"

By now she should have guessed.

"Happy birthday!"

You can do this trick for any birth date, if you give it a bit of thought. Try it and see.

The Secret: The trick is simply to make up ahead of time a problem that will result in 21 or any other date you want to show. If you start with 50 − 30 (there are 30 days in June), all you have to do is add 1 to get 21. When you plan the trick, remember how many days are in the month. If a birthday is July 22, subtract 31 from 50 and add 3.

CUCKOO, CUCKOO

A little wooden man stands on top of a cuckoo clock. Each time he hears the clock strike once and the bird call cuckoo, he jumps a certain number of times—twice for two o'clock, three times for three o'clock, and so on. How many chimes does the little man hear in 24 hours?

Think a minute before you look for the answer.

SHORTCUT

A German boy named Karl Friedrich Gauss grew up to be one of the world's greatest mathematicians. When he was six years old, his teacher told him to add up all the numbers from 1 through 100. He did this so quickly that his teacher thought he was cheating. But Gauss had simply figured out a shortcut in calculation. Can you figure it out?

The Secret: Instead of adding up a long column of 100 numbers, Gauss added 1 to 100 and multiplied the sum by 50.

You can see how this works out if you start with an easier problem: Add up all the numbers from 1 through 10. First, write the numbers in two columns in the order shown below. Then add them across, and then add up the sums in this way:

$$
\begin{array}{r}
1 + 10 = 11 \\
2 + \ \ 9 = 11 \\
3 + \ \ 8 = 11 \\
4 + \ \ 7 = 11 \\
\underline{5 + \ \ 6 = 11} \\
55
\end{array}
$$

As you can see, there are 5 pairs of numbers, and each pair adds up to 11. So, instead of adding the five 11s, you multiply: $11 \times 5 = 55$.

If you write the numbers 1 through 100 the same way you will have 50 pairs, each pair adding up to 101. Gauss saw this and got an answer: $101 \times 50 = 5{,}050$. Check it with your calculator.

Cuckoo Answer: Since when can a wooden man hear?

SNEAKY RACE

This is a tricky game for you to play with two people who also have calculators. It is most likely to puzzle them if they have never heard of Fibonacci (pronounced Fib-oh-NAH-chee) numbers. Even if they have, they may not suspect a plot when you suggest the game.

"I'll bet that my calculator is faster than either of yours," you say to Lucy and Pete. "Let's test them on this fancy addition problem that my dad gave me this morning. We'll write it down first."

All three of you copy this column of numbers on paper:

"We don't have to do the whole problem," you say. "Let's just add up any ten numbers in the column. Lucy, you choose which one we start with, and then we'll draw a line under the tenth number to be added."

Suppose Lucy chooses 8. All of you draw a line under 610.

"Now, when Lucy gives us the signal," you propose, "we'll all start adding those ten numbers. The calculator that gets the answer first is the winner."

"Go!" says Lucy.

She and Pete begin to press buttons. So do you, but you don't add anything. You simply look at the sixth number below 8 in the column and multiply it by 11: 144 × 11 = 1,584. You wait a

There are other intriguing facts about Fibonacci numbers. Here is one to try: Choose any three consecutive numbers in the column. Suppose you choose 3, 5, and 8. Multiply the middle one by itself: 5 × 5 = 25. Now multiply the first number you chose by the third: 3 × 8 = 24. Compare the two results. The difference between them is 1. You will find that no matter which three consecutive numbers you choose for the experiment, the difference is always 1.

moment or two, then quietly lay down your calculator and sit back. Naturally, Lucy and Pete notice that you finished before they did, but they can hardly believe it when they see that the answer in your display is the same as theirs. If they protest that you cheated somehow, remind them that you didn't choose where to start the addition. Lucy did. If Pete wants to choose a different group of consecutive numbers, you are happy to have another three-calculator race.

The Secret: This trick works no matter what number you start with. It was discovered almost 700 years ago by Leonardo of Pisa who was called Fibonacci (*Fibonacci* means something like "good-natured kid"). The Fibonacci numbers in the column are arrived at in a special way. The sum of the first two numbers gives you the third. The sum of the second and third gives you the fourth. Third plus fourth gives the fifth, and so on.

You can make this game even sneakier. Just write up your own special sequence of Fibonacci numbers. For example, 11, 12, 23, 35, 58, and so on. This sequence will look like just any other big addition problem, and your friends won't suspect the trick

If you want to find out exactly why multiplying by 11 works, ask your math teacher—or wait until you can read about the trick in an algebra book.

EVEN SNEAKIER

"You still don't believe my calculator can beat yours?" you say to Lucy and Pete. "Let's try again. This time we'll add up all the numbers between 4 and 30. Let's see who gets the answer first."

"All right. Go!" says Lucy.

They push buttons madly. Again, you just pretend to be working hard. In a few moments you press "clear" and then do the following calculations:

1. **Add the first and last numbers in the series: 4 + 30.**
2. **Divide by 2: 34 ÷ 2 = 17.**
3. **Multiply by the total number of numbers in the series: 17 × 27 = 459. (The thing to remember here is that from 4 through 30 there are 27 numbers, *not* 26 as you might think.)**

Again you put down your calculator and lean back while Lucy and Pete are working.

You've won again—but with smart math and not a super calculator.

The Secret: The easiest way to see how this trick works is to try it with five smaller numbers: 2 + 3 + 4 + 5 + 6 = 20

1. **Think of each of these numbers as a pile of blocks—2 blocks in the first pile, 3 blocks in the second, and so on. You want to find out how many blocks there are in the whole string. This is what they look like:**

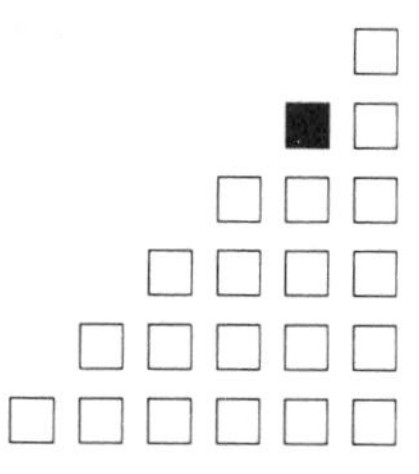

2. **Look now at the first and last piles of blocks. Add them together: 2 + 6 = 8. Next look at the middle pile. It has 4 blocks, just half as many as the two end piles added together: 8 ÷ 2 = 4.**
3. **Now suppose you borrow blocks from the 4th pile and the 5th pile. If you move them, you can make all the piles equal: 5 piles with 4 blocks in each: 4 × 5 = 20.**

By following the three steps, you can quickly find the sum of any string of consecutive numbers.

BAD BET

At the track meet between Lincoln School and Jefferson School, Dick was standing next to Paddy, a kid from Jefferson.

"I'll bet Lincoln is going to win the broad jump," Dick said.

"How much do you bet?" Paddy asked.

Dick pulled out all the change he had in his pocket. "I'll bet half of this," he said.

"No," said Paddy. "I want half plus one cent more."

Dick agreed. Unfortunately, Jefferson won the broad jump. Dick paid his bet.

"Well, I'll bet half of what I have left that Lincoln wins the high jump," Dick said.

"No. Half plus one cent more."

Again Dick agreed, and again Lincoln lost. But Dick remained loyal. He made yet another bet that Lincoln would win the shot put. Alas! Jefferson had a man with muscles like hamburger buns. Dick paid his bet, but he was still hopeful. Making the same bet twice more—on the pole vault and on the hundred-yard dash—Dick lost each time. After paying Paddy, Dick had exactly one cent left.

How much change did Dick have in his pocket at the beginning of the track meet?

Here's how to find out: Figure the answer by going backwards. Starting with the penny left in Dick's pocket, add 1 for the last cent that he gave Paddy. The least he could have given Paddy was 3 cents, so he had 3 cents + 1 cent or 4 cents to bet on the hundred-yard dash. Then he must have had 4 cents plus 1 cent multiplied by 2, or 10 cents, before the pole vault. Before the shot put, he must have had 10 cents plus 1 cent multiplied by 2, or 22 cents. Before the high jump, he must have had 22 cents plus 1 cent multiplied by 2, or 46 cents, to bet on the broad jump. Just to make sure you're right, use your calculator to do Dick's problem forwards.

DINO-MATH

This is a fun trick to play on someone who is a dinosaur fan. The artist has drawn pictures of six meat-eating dinosaurs and six plant-eating dinosaurs. It is easy to tell them apart. All of the meat eaters are standing on their hind legs. The plant eaters are on all four legs. Some are big and some are small, just as they really were millions of years ago.

"It's a funny thing about my calculator," you tell your friend. "It can't keep a secret. Let's see if you can hide some numbers from it."

You hand him a pencil and paper. Then tell him to write down the names of several pictured dinos on the paper—as many of each kind as he wants. You won't look at what he has written.

When he has finished, hand him your calculator and ask him

to enter some numbers. Three of the numbers will be secret.

1. **The first secret number is the number of meat eaters he has written down. (Suppose it is 2.)**
2. **Multiply by 2: $2 \times 2 = 4$.**
3. **Add 3: $4 + 3 = 7$.**
4. **Multiply by 5: $7 \times 5 = 35$.**
5. **Add a second secret number, which is the number of plant eaters he has written down. (Suppose it is 1. Then $35 + 1 = 36$.)**
6. **Multiply by 10: $36 \times 10 = 360$.**
7. **Add the third secret number, which is the number of little dinosaurs he wrote. (Suppose it is 1. Then $360 + 1 = 361$.)**

Now ask to have the calculator back. Quickly you subtract 150. This leaves 211 in the display.

"You've written down the names of 2 meat eaters and 1 plant eater," you announce. "And only one of the creatures is small. Looks like my calculator still can't keep a secret."

The Secret: Look at what you have done in steps 2, 4, and 6. You doubled the first secret number, multiplied it by 5 and then by 10. This pushed the secret number into the hundreds place. In steps 4 and 6 you pushed the second secret number into the tens place. In step 7 you get the third secret number put into the ones place.

Now where does 150 come from? In steps 3, 4, and 5, you have put into the calculator an extra number—3—and multiplied it by 5 and 10 in steps 4 and 6: $3 \times 5 \times 10 = 150$. So when you subtract that 150, only the three secret numbers remain in the display.

Quizzle: Wesley's grandma is a sly old lady. One day she said: "Wesley, how many months have 28 days?"

"One, of course," Wesley snapped back. "February."

Do you think Wesley was right? Find out on page 36.

TATTLETALE

"I have to apologize for my calculator," you announce. "It's getting to be a real tattletale—can't keep a secret."

Jack, Sue, and Joe look doubtful.

You hand the calculator to Jack and say, "Here—let's test it. Enter a secret number less than 10 and remember what it is."

Jack presses the buttons, and you give him more instructions:

1. Multiply by 2.

2. Add 3.

3. Multiply by 5.

4. Add 7.

Now you take the calculator and hand it to Sue.

"Sue, don't clear the display. Just add your own secret number to the one that's already there. Remember your number. Okay? Now:"

1. Multiply by 2.

2. Add 3.

3. Multiply by 5.

Next it is Joe's turn. You tell him simply to add his own secret number. Then you take the calculator back and talk to it: "Can you keep a secret now?

You press buttons to subtract 235 from the number in the display and then pretend to listen. Finally, you shake your head and say, "I'm afraid this machine is a blabber mouth."

You show the display. "Jack, the first digit on the left is your secret number. The middle digit is Sue's. The one on the right is Joe's."

The Secret: Suppose Jack picked 4 as his number; Sue chose 7; and Joe, 8. Work the problem and you will see that after you subtracted 235, the final display is 478. The instructions you gave for multiplication put Jack's secret number in the hundreds place and Sue's in the tens place. Joe's number was last, so it fell in the ones place. All the other additions just brought their total up to 235, which you subtracted and so ended with 478.

Nobody knows who invented the preceding trick, but it has been around ever since a French mathematician named Claude Gaspar Bachet wrote about it in the year 1642. At that time, a lot of people were studying math because many kinds of businesses were starting and growing, and employers needed skillful bookkeepers.

Quizzle Answer: All months have 28 days.

SURPRISE YOURSELF

It's not easy to surprise yourself, but here's a way to do it.

1. **Enter a number between 1 and 10. (Remember what it is.)**
2. **Multiply by 8.**
3. **Add 12.**
4. **Divide by 4.**
5. **Subtract 3.**
6. **Divide by 2.**
7. **Press the "=" button.**

The answer in the display is the number you started with! That will happen no matter what number you enter first. Why?

The Secret: Here are some pictures that will explain this and many other games and puzzles from now on.

Pretend that your original number is a pod full of peas. All the other numbers that you add or subtract in the problem are peas outside the pod.

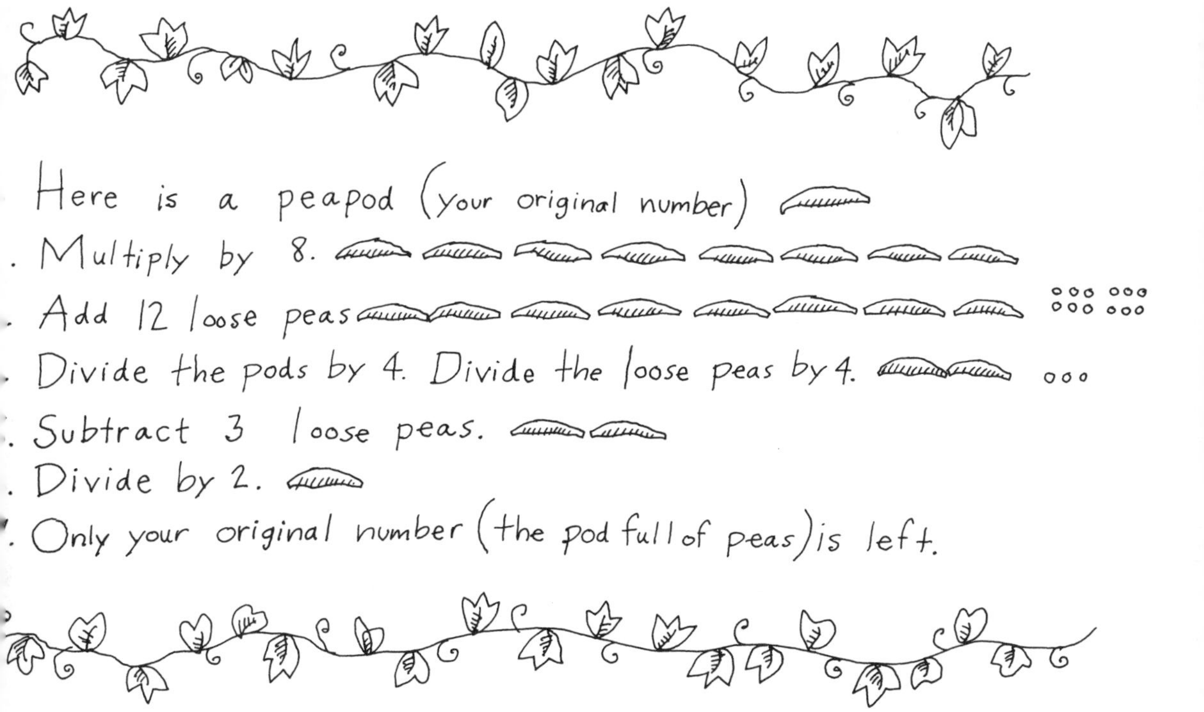

BEWITCHED

Know any superstitious people? Some folks believe that thirteen is an unlucky number, and that Friday the thirteenth is an especially unlucky day. If you look at a calendar and it's Friday the thirteenth, you can play this trick on a friend.

Suppose your friend's name is Jody. You remind Jody of the old superstition about the number thirteen. Then you say, "By the way, my calculator has been taking lessons from a witch. This morning she told me that because today is the thirteenth, this wicked machine can put a hex on other calculators. Do you want to see if it really is that powerful?"

1. **You turn on your calculator and tell Jody to enter a number—any number.**
2. **Add 16.**
3. **Multiply by 2.**
4. **Add 4.**
5. **Divide by 2.**
6. **Subtract the number you entered in step 1.**
7. **Subtract 5.**

Sure enough, the display shows 13. Then let Jody try it with a different number in her own calculator. The result will always be 13. Can you work out the peas-in-a-pod test?

The Secret: The peas-in-a-pod test looks like this:

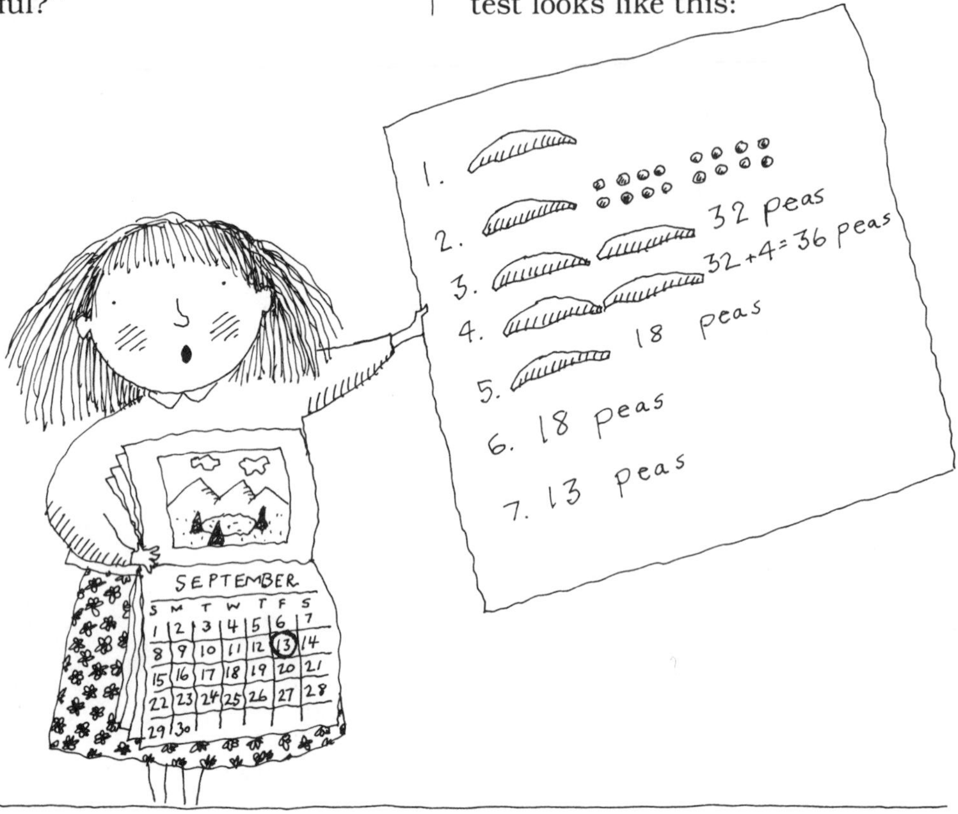

STILL BEWITCHED

"Hmmm," you say to Jody. "Let's try again."

These are the new instructions:

1. **Enter a three-digit number. Remember it.**
2. **Repeat the number. That is, if your number is 133, press 133 again.**
3. **Divide by 7.**
4. **Divide by 11.**
5. **Divide by your secret three-digit number.**

Again, the number in the display is 13.

The Secret: This time the hexing trick depends on the peculiar qualities of the number 1,001.

Multiply any three-digit number by 1,001. The answer will be the 3 numbers repeated side by side. For example, 133,133.

Multiply $7 \times 11 \times 13$ and you get 1,001.

In steps 1 and 2 you have multiplied the secret number by 1,001.

In steps 3, 4, and 5 you have got rid of the secret number and have reduced 1,001 first by 7 and then by 11. That leaves 13.

Here is a puzzle that teachers gave to arithmetic students a thousand years ago:

A boy was sent to the market with 100 coins to buy some animals. In those days a camel cost 5 coins. A donkey cost 1 coin, and you could buy 20 sheep for 1 coin. How many of each did the boy come home with?

Hint: More than one answer is possible. Start with camels because they cost the most:

16 camels $\times$ 5 coins each = 80 coins.

Next try adding 15 donkeys:

15 $\times$ 1 coin each = 15 coins.

The boy has now spent 80 coins + 15 coins = 95 coins.

That leaves 5 coins for sheep. He can buy 20 sheep with 1 coin, so 5 coins $\times$ 20 = 100 sheep.

Now figure out how many animals a dopey boy could get if he started with 20 camels.

TRICKY TALK

"Hi!" you say to Pablo. "Did you know that my calculator can carry on a conversation with yours? You don't believe me? Let's try."

You explain that you will start the conversation by giving Pablo's calculator a problem. First, he is to enter a number—any number. It can be a big one—four digits if he likes. To keep track of the number, he should write it on a piece of paper.

"But don't show it to me," you say. Then you give him these instructions:

1. **Subtract 1 from the number entered.**
2. **Multiply by 2.**
3. **Subtract 1.**
4. **Add the original number. (If he has forgotten it, he can refer to the note he made to begin with.)**

"Good," you say. "Now tell me the answer that's in your display, and I will enter it in my calculator. That will start the conversation between your calculator and mine."

1. **You enter what Pablo tells you.**
2. **Add 3.**
3. **Divide by 3.**

"Well," you say, "my calculator was asking yours what your original number was. And this is your calculator's answer."

Pablo looks at your display. There, indeed, is his original number!

The Secret: Suppose Pablo started with 4,328. This is what the crafty calculation does:

4,328	Here you have
− 1	subtracted 1.
4,327	Here you double that
× 2	subtracted 1, so it is
8,654	now 2.
− 1	You subtract 1 again,
8,653	making a total of 3.
+ 4,328	By adding this you
12,981	have put 3 times the original number, minus 3, into Pablo's display.

So you add the missing 3, divide by 3, and there it is: 4,328—the original number.

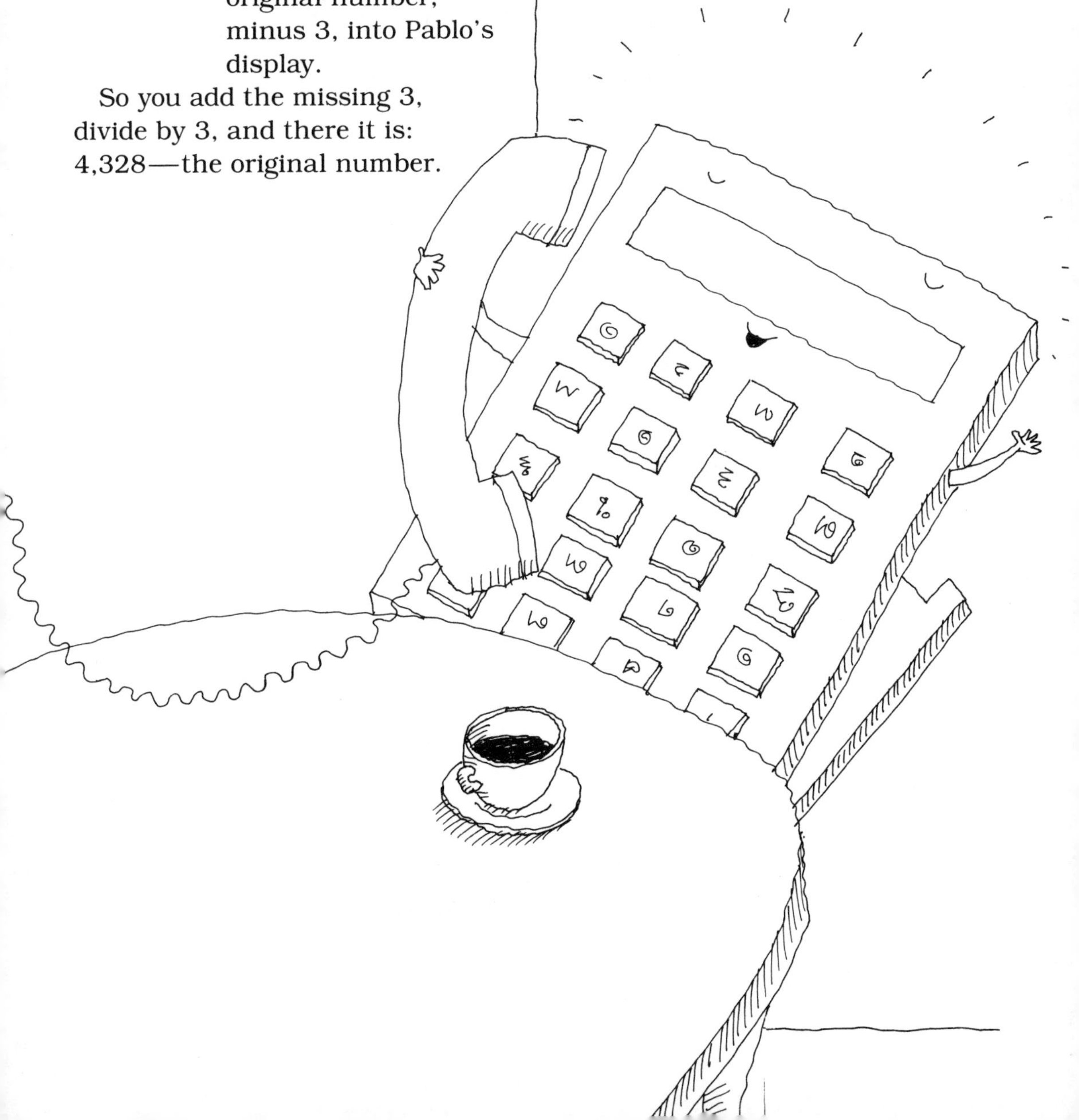

Mind Reading with Your Calculator

"My calculator has just graduated from Mind-Reading School," you announce. "It now has power to perform astonishing feats."

You can try your mind-reading tricks on one person at a time. Or you can get a friend to help you put on a show for a group of people. You will really surprise them!

BRAINSTORM

Suppose Maya agrees to be your partner. The two of you practice the routines ahead of time. There are a lot of steps you will want to memorize. That may be hard, but Maya can be your coach. If there are some steps you're not quite sure of, don't take a chance. Write them on cards that Maya will keep hidden until just the right moment. Then she will hand you one and pretend it is a message from someone in the audience. You look at the card and say, "It says here that Martin is too smart to let any old machine read his mind. Come on up, Martin."

You hand your calculator to Martin and turn your back. "Ready?" you say. "Think of a number, but don't tell me what it is."

Here Maya gives Martin a pencil and a slip of paper. "Write down your secret number," she says, "so you can use it later."

Now you dictate the problem.

1. **Enter your secret number.**
2. **Add 2.**
3. **Multiply by 3.**
4. **Add 4.**
5. **Subtract your secret number.**
6. **Subtract 4.**
7. **Divide by 2.**
8. **Subtract your secret number again.**

"All right," you say. "That's a big enough problem. Now, Martin, look at the answer, but don't tell me what it is. Just concentrate on it so that my calculator can transmit the answer to me."

Ask the audience and Maya to concentrate on the number, too. You shut your eyes and frown. Then, "I'm getting something! It's coming . . . it's coming! The answer is 3."

If Martin has pressed the right buttons, the number is certainly 3. But suppose his finger slipped, and the answer came out wrong. *Don't blame Martin.* That would spoil your show. Instead, scold your calculator! The naughty little rascal has tried to make you look foolish, and you want it to be more careful next time.

If Martin did the operations correctly, you say, "Maybe you think my little machine isn't really smart, and I just made a lucky guess. So let's try again." Then on you go to another trick.

The Secret: By now you know the secret of the peas in a pod. You can use it to make up other mind-reading tricks of your own. Just put and take your own pods and loose peas in a problem. Remember that your plan is always the same: Your instructions undo each other. If you add or subtract, multiply or divide in one step, you subtract or add, divide or multiply later on.

FAST AS LIGHTNING

"This time, Martin," you say, "my calculator is getting more confident. It's picking up speed. Let's go!"

Your instructions are:

1. **Enter any three-digit number. Be sure to remember it.**
2. **Multiply by 10.**
3. **Subtract your secret number.**
4. **Divide by your secret number.**
5. **Multiply by the number you see in the display.**
6. **Add 19.**

Here you stagger and hold your head, as if something has hit you.

"I've got it already!" you cry. "It worked like lightning! It's beautiful—my favorite number! Exactly 100!"

The Secret: The secret is in the first four steps. No matter what three-digit number is entered in step 1, the number at the end of step 4 is always 9. Now you can be sure that steps 5 and 6 ($9 \times 9 + 19$) will produce 100.

EVEN FASTER

For your next trick Maya gets Connie to volunteer because you know Connie is especially good with a calculator.

"Please concentrate on the answer to the problem that I'll dictate," you tell Connie, "and my calculator will flash the answer from your brain to mine as fast as *greased* lightning."

Here are the instructions you give Connie:

1. **Enter the year you were born.**
2. **Add the year when something very important happened to you. (If Connie plays basketball, you suggest it could be the year she made the team. Or the year she went to camp or broke her leg. She is not to tell you what the year is.)**
3. **Add the age you will be at the end of the present year.**

At this point you exclaim: "Hold everything! I can already feel this will be a big number. I need paper and pencil to be sure I don't copy it wrong."

Maya hands you a pad and pencil, and you go on:

4. "Connie, add the number of years that have passed since the year when that important event happened. Now *concentrate.*"

Immediately you smack your forehead and say, *"I've got it!"*

You pretend to scribble on the pad, then show the answer: It is exactly the number that appears in the display.

The Secret: You have written down the number ahead of time. The scribbling is just for stage effect—as if the number is flashing from the calculator into your brain. The number is the current year—for example, suppose it is 1992—multiplied by 2, which makes 3,984. Can you figure out why?

Look at it this way: Two of the numbers in your instructions to Connie—the year she was born and her age at the end of this year—add up to 1992. If Connie broke her leg in 1989, that number, plus the 3 years that have passed since then, makes another 1992. Add the two together and you have the number in the display.

When Houdini, the great magician, was young, he and his wife Beatrice entertained audiences with mind-reading tricks. Eventually, however, they came to dislike the business of tricking people who took mind reading seriously. After that, Houdini always explained how he did his stunts. He also spent a lot of time exposing the hoaxes of others who took advantage of audiences.

TELEPHONY NUMBERS

You and Maya must plan this trick ahead of time. First, you make sure that a certain person—let's say Ben—will be in the audience. Then you make sure you know his telephone number. Suppose it is 459-3262. You subtract that number from 8,100,000, making 3,506,738. (You will find out in step 8 below the reason for using 8,100,000.) Either memorize 3,506,738 or write it on a small slip of paper that you can take out of your pocket and hide in your hand when you need it.

After Maya gets Ben to volunteer, you say, "As you know, my calculator has extraordinary powers. Not only can it read your mind, it can influence your thoughts."

These are your instructions:

1. **Enter a secret three-digit number.**
2. **Multiply by 10.**
3. **Subtract the secret number.**
4. **Divide by the secret number. (The answer in the display will be 9. You can count on that.)**
5. **Multiply by 9: $9 \times 9 = 81$.**
6. **Multiply by 10: $81 \times 10 = 810$.**
7. **Multiply by 10,000: $810 \times 10{,}000 = 8{,}100{,}000$.**
8. **Subtract 3,506,738: $8{,}100{,}000 - 3{,}506{,}738 = 4{,}593{,}262$.**

You pause for a few seconds. "Examine the answer in the display," you say, looking suspicious. "Are you thinking about that number for a special reason? It must mean something to you—yes, yes! Now I'm getting a message. The answer in the display is *your phone number!*"

You just hope that Ben doesn't say, "Ha, ha! My phone number has been changed."

But if he makes a mistake, and the wrong number comes up in the display, just say your calculator likes to tease people. Then quickly go on to your next amazing feat.

MENTAL MASTERY

This is a good trick to use if there happens to be a grownup in the audience, and if Maya can persuade him or her to volunteer.

"Maybe you doubt that my calculator is a mind reader," you say. "But I assure you it has gone to school with a real Master. With your help, we can test its skill."

These are the instructions you give to the grownup:

1. **Enter your age.**
2. **Multiply by 2.**
3. **Add 5.**
4. **Multiply by 50.**
5. **Subtract 365.**

Here you pause and ask: "Do you have some change in your pocket? You can make a guess at how much it is if you don't want to look. Just choose any amount under a dollar."

The instructions continue:

6. **Add the amount of change. But remember how much it is—not more than 99 cents.**
7. **Now add 115.**

You reach for the calculator. "Let's see if it's sending me a message—something that you know. Yes! That's it. Your age is . . . ."

You read the first two digits in the display, and sure enough they give his or her age. You look at the last two digits in the display and announce the exact amount of the person's change.

The Secret: To figure out how this stunt works, think about the first two digits. They are in the thousands and hundreds places. To get the person's age into those places you multiply the age by 2 and then by 50. In between the two steps you add 5, just to make things more confusing. That 5 has now been multiplied by 50, making 250. To get rid of it you add 115 and subtract 365. What remains now is only the person's age multiplied by 100. Suppose he or she is 21 years old. The display will show 2,100. Now all you need to have the person add is the change in his or her pocket. Suppose it is 67 cents. *Presto*—2,167!

FARE AT THE FAIR

For a change of pace you can sit down and use a different tone of voice, as if you are being pals with the audience: "You know, I have a friend named Bernardo who runs the roller-coaster at the fair every year. He has a calculator that is just as smart as mine is. Let me tell you about it."

Your story can go something like this:

"First, I should say that Bernardo is a real friend of little kids. He lets you have a ride for half price if you are under ten years old. But sometimes older kids pretend they are not yet ten. If Bernardo suspects somebody is trying to fool him, he gets out his Age Detector. That's his smart calculator. Now Bernardo says to one boy, 'Before I sell you a ticket, I want you to do a little experiment for me.'

"The kid thinks this may be fun and agrees. He presses the calculator buttons that Bernardo tells him to press.

"When Bernardo completes the last step in the problem, he looks at the number in the display and says 'Ha! Just as I thought. My calculator is an Age Detector. It tells me you are eleven years old.' He then tells the boy the exact month and day of his birthday.

"But the kid interrupts Bernardo, saying, 'I know, I know. This is the way your Age Detector works.' And he astonishes Bernardo with his explanation.

"Do you know what? My friend Bernardo gave that smart kid a *free* ride."

Now you stand up and say in a businesslike voice, "Bernardo's calculator taught mine how to be an Age Detector, too. Who would like to test it?"

Suppose Millie volunteers. Before you begin, Maya makes sure that Millie knows which month of the year is first, second, and so on. That is, January is number 1, February is number 2, and so on.

Now you tell Millie what to do:

1. Enter the number of the month in which she was born.

2. Multiply that number by 100.

3. Add the day of the month in which she was born.

4. Multiply by 2.

5. Add 7.

6. Multiply by 5.

7. Add 4.

8. Multiply by 10.

9. Subtract 390.

10. Add her age.

"Enough," you say. "In a minute you'll see how powerful my Age Detector is."

Suppose the number in the display is 102,111.

"I'm glad to find this out," you announce. "You are eleven years old. Your birthday is October 21."

The Secret: To get that secret information you read the numbers in the display from right to left. The first two digits on the right are Millie's age. The two digits on the left are the number of her birth month. The middle digits are the day of the month. Try it with your own birthday to see how it works.

You can see that the plan for this stunt is the same as the plan for *Mental Mastery* on page 47. The difference is that you multiply to get the month and the day into the hundred thousands, the ten thousands, the thousands, and the hundreds places, respectively. Step 9 gets rid of the numbers added in steps 5 and 7. Step 10 adds the person's age.

Watch out for a smart-aleck kid who is really ten years old but says he is six and was born on January 2. The display will show only five digits (10,206). Don't panic. Reading from right to left you get 6 and that's the right age. The zero in front of it means that there is no number in the tens place. The zero in front of 2 is just holding the thousands place. And the 1 stands for January. If the kid's birthday was in October, the tenth month, the display would show 100,206.

MEET THE PSYCHIC

Save this trick for last when you are performing for a group. It could be your claim to fame! Later, you can try it on your teacher, too.

"I can show you an important mind-reading test," you tell Mario. "I'm going to test my calculator's power to transmit a sentence found in one of my favorite books. While you read the passage to yourself, my calculator will magically whisk the words to my mind from yours. Then I'll repeat them just as they are in the book."

Of course, this boast means that you have brought the book with you and have already memorized a certain sentence in it. Memorize a whole paragraph, and your career is made.

1. **You hand the calculator to Mario and say: "I want you to enter a three-digit number. But first I have to warn you that my little mind reader is being fussy today. It wants to digest big digits before smaller ones. So, please start with a number in which the first digit is larger than the third digit—such as 632. Got it? Good.**
2. **"Enter the number. Now press the minus button and reverse the number. That is, enter the**

last digit, then the middle digit, then the first digit. Now push the equals sign."

3. **You pause again. "I think my calculator is trying to tell me something. Is the number in the display a three-digit number?"**
4. **If the answer is "yes," you say: "Good. Now press the add button. Enter the reverse of the number in the display. Press the equals button. Now we are ready for the extraordinary test." (If the answer is "no," you'll see in a moment what to do.)**
5. **You give Mario the book you have brought. Let's suppose it is *Charlotte's Web.* "Look at the first three digits of the number in the display. Don't tell me what they are. Now turn to that page in the book."**
6. **You wait till Mario finds the page. "Now look at the fourth digit in the display. Don't tell me what it is. Just count down that number of lines on the page and read the sentence to yourself."**
7. **You pretend to be concentrating. In a moment or two you say triumphantly: " 'Well, do you understand it?' asked Mrs. Arable."**

You already know that the sentence is the ninth line on page 108 in *Charlotte's Web.* For really big applause, also memorize the next four lines on that page!

This trick works with any other book you want to use. Just look up line 9 on page 108 and memorize it.

The Secret: You have buried your experiment in a mathematical fact older than the first arithmetic book. No matter what three-digit number Mario chooses, he is bound to get either 396, 198, or 99 at the end of step 2. Now look back at step 4. What if Mario says the display does *not* show a three-digit number? That means the display shows the number 99. Keep your cool. You say, "Ah, yes. My calculator is hungry. We'll feed it two more numbers. Add 495. Now add 495 again." That brings the number in the display up to 1,089. And you go on from there with the instructions.

If you want to use a book other than *Charlotte's Web,* you may have to hunt a while until you find one that has just the right line 9 on page 108. Discouraged? Then try this: Look at page 109, line 2. If you like that passage, just put one more step in your trick. Tell the volunteer to add 3 as your last instruction. That means 1092 will show in the display. Try other additions, too, until you find just the right passage to memorize.

P.S.: Your teacher will like it if your mind-reading machine produces a verse or two from a book of poetry.

UPSIDE DOWNERS

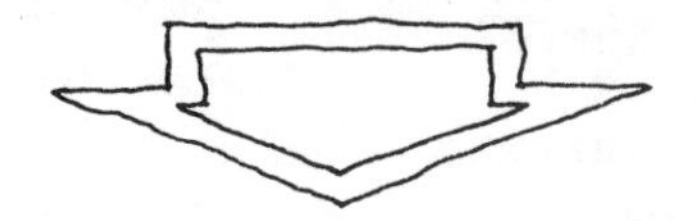

Try this on your calculator. Push the buttons to make the number 0.7734 in the display. Now turn the calculator upside down. What do you see? Some of the numbers are letters when you stand the display on its head, and they look like hello. Here is a story with some upside downers for you to calculate.

BIG DEAL

Jesse bought a house for $9,768. He put on a new roof that cost $2,500 and built a swimming pool for $4,600. When it was all done, a man came along and said, "I'd like to buy this place. I'll give you $16,500 for it."

That seemed like a lot of money to Jesse, but just to be sure he asked his friend Luke what to do.

Luke said, "1,000 + 105 × 7."

That's just what Jesse did.

When Jesse's daughter Rosie came home from school, Jesse told her what a good deal he had made. Rosie took out her calculator and did some adding and subtracting. When she finished, she said to her father, "Poor Daddy! I think you are a big 16,868 + 18,141!"

"Why?" Jesse asked.

"Because you sold the house at a 5,139 + 368." Rosie answered.

Can you tell why Rosie used the numbers 16,868 and 368 when she scolded her father?

Answer: Rosie added how much her father had spent on the house and subtracted the amount that he was paid for it (16,868 − 16,500 = 368). She found that he had lost $368.

1 = I
3 = E
4 = h
5 = S
6 = g
7 = L
8 = B
9 = G
0 = O

THE GORILLA CHOW MYSTERY

Timid Tim worked at the zoo. His boss Ellsworth Hamilton was in charge of buying food for the animals. Recently, the president of the zoo had suspected that Hamilton was not reporting honestly the amount that the animals ate. In fact, the cost of their food seemed to be very high. So the zoo president hired a private investigator to do some detective work.

The investigator decided to visit the animal cages first. There he found Tim giving chow to the gorillas.

"You work for Mr. Hamilton?" the investigator asked.

Tim scowled.

"What's the matter?" the investigator asked. He didn't know that Tim was very timid, and that he was especially scared of his boss. "You know Mr. Hamilton, don't you?" said the detective.

Tim kept on scowling. Then suddenly he took out a pencil and wrote this on a piece of paper that was lying on the ground: 39,491,229 + 41,260,305.

The investigator didn't know that Tim was really a math whiz kid. Tim had a summer job working at the zoo because he wanted to make enough money to go to college.

The investigator was puzzled by the numbers Tim had written.

But because he was curious, he took out his calculator and did the addition. The answer in the display was 80,751,534. For a moment he stared at it and then gave Tim a big smile.

"So you don't like your boss. You think *he's 1 slob*," he said. "Tell me more."

Tim felt a little braver now. So he wrote this: 309,326 ÷ 422

"Hmmm," said the investigator after he did the division. "So—!Hamilton is slippery as an *eel!* Is that what you mean?"

Tim nodded vigorously.

While the investigator was wondering what to do next, Tim went on dishing out gorilla chow. That gave the investigator an idea. He reached in his briefcase and pulled out a paper showing how much money Hamilton had charged the zoo for gorilla and monkey chow and hay for the zebras.

Tim studied the figures. Then he wrote this: 813 − 496.

"Aha!" said the investigator after he did the subtraction. "Hamilton has told a *lie*. Can we prove it?"

Tim hesitated and looked around suspiciously. He quickly wrote down some numbers, then hurried away.

This time the inspector was really puzzled by the answer to Tim's problem: 7.15 × .10 = .715

"Silo," he said to himself. "That's a big thing like a tower on a farm." Finally, he had an inspiration. He checked Ellsworth Hamilton's address and solved the mystery. Hamilton lived outside the city on a farm that had a silo and a barn. The barn was full of hay that Hamilton had charged to the zoo but was feeding to his own cows. The silo was packed with sacks of gorilla and monkey chow that he fed to his pigs. No wonder the zoo animals seemed to eat so much!

Hamilton suspected he was being investigated and decided to disappear. But the police caught up with him in 0.140.

P.S. Tim kept his summer job, went to college, and now teaches math. He is still a little bit timid.

HA-HA!

"Want to hear a joke?" Jeffrey asked Angela.

"Sure," she answered.

"The math teacher gave Greg this problem: If you found a quarter in one pocket and a dime in the other pocket, what would you have? Greg answered: 'Somebody else's pants.'"

"I think that joke is 317,700 + 15," Angela told Jeffrey.

Jeffrey is usually very polite. But when Angela criticized his joke his feelings were hurt, and he said, "I say that you're a 2,002 × 4."

"Why?" Angela asked.

"You don't know how to spell, *silly.*"

SMART ANSWER

"My calculator is so smart that it knows the answer without ever doing the problem," you tell Jason.

"Prove it," says Jason.

"Okay. We'll both enter 169 in our calculators."

When 169 is in both displays, you say, "Now add 522."

Jason does so, but he notices that you do not move a finger.

"Show me the answer," you say. "That's right—691."

"Show me yours," Jason demands.

You turn your calculator upside down and there it is—691.

"I told you it could get the answer without hearing the problem!"

QUICKIES

Question: Why did a whole flock of penguins decide to rob a bank?

Answer: Because they had so many big 45,618 + 12,100.

Question: What do hotels and Congress have in common?

Answer: 3,758,415 + 1,560,392

Question: What did the snake say to the Wizard?

Answer: 5,000 + 514.

Question: What did the Wizard say to the snake?

Answer: 1.59 − 1.51.

Question: What did the man say when he found out his horse had won the Kentucky Derby?

Answer: .8080808 ÷ 2

Question: What does Robinson Crusoe signal to a passing ship?

Answer: $16 \times 16 \times 2 - 7$

Question: What does he do when the ship doesn't stop?

Answer: $290{,}267 \times 2$

Question: If Jimmy went to the ice-cream store and ate six double dishes of maple walnut, what would his friends call him?

Answer: A $285 + 619$.

Question: If he did that every day for two years, what would they call him?

Answer: $95{,}072 - 59{,}692$.

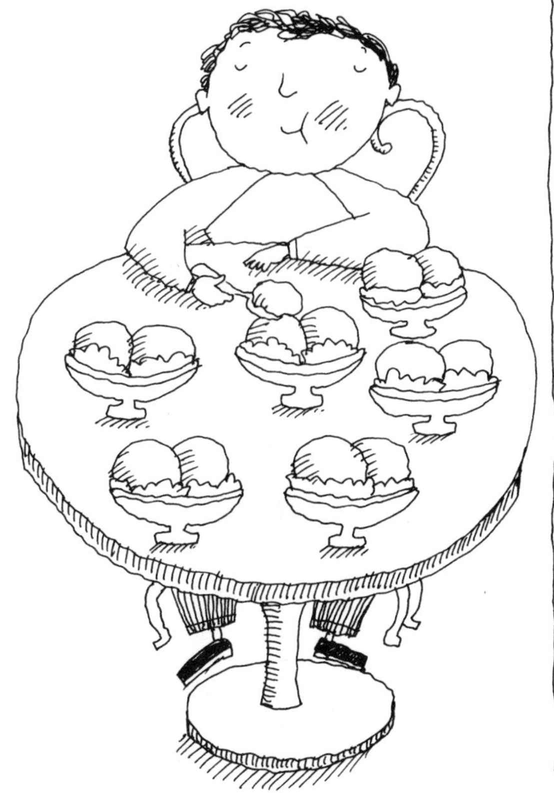

HOW PATSY CAUGHT THE CROOK

Patsy's mother and father had to make a business trip to New York during spring vacation, and Patsy came with them. They stayed in a hotel that had a big lobby, where Patsy sometimes sat when her parents were busy. Sometimes she read a newspaper or book. Sometimes she brought her calculator along and did problems for fun. And sometimes she just watched the many different people who also sat in the hotel lobby.

One day, Patsy noticed a man who was working with a calculator at a small table not far from her. He seemed to be doing some complicated problems, and Patsy began to wonder. Was he taking care of some sort of business? Maybe he was figuring out how much he had spent that day? Every once in a while he wrote something down on a piece of paper. Finally, he took out a clean piece of paper and copied down what he had written. Then he got up and hurried away.

Patsy was curious. She had noticed that every once in a while he turned his calculator upside down, studied it, and scribbled something. What did that mean? It was certainly strange behavior, Patsy thought, as she got up and started back to her room. On her way to the elevator something caught her eye. On the table where the man had been sitting was the sheet of paper on which he had scribbled.

Patsy hesitated. Clearly, the man had finished copying his notes. But had he meant to leave this paper behind? She *did* want to see what was on that paper. Feeling guilty, she snatched it up, put it in her sweater pocket, and hurried to the elevator.

When Patsy's mother and father came back from a business conference, they found her so excited she could hardly talk. She had been working with her calculator, she said, and showed them the sheet on which the man had written a final set of numbers. It looked like this:

3,475,509 + 4,263,299

238,931 + 262,803 $5,000

9,359,498 − 3,825,461

6,901 − 1,720 $3,500

10,157,849 − 2,443,676

3,719 − 639 $1,650

"I did the calculations," Patsy said. "And this is what I got."

7,738,808	**501,734**
5,534,037	**5,181**
7,714,173	**3,080**

She showed the numbers to her parents. They looked puzzled for a minute. "Turn them upside down," said Patsy. So her mother entered the numbers and each time turned the calculator upside down.

This is what she saw:

Bob Bell	**Helios**
Leo Hess	**Ibis**
Eli Hill	**Oboe**

"Patsy," said her father, "I don't see why you are so excited."

"That's because you've been too busy to read the newspapers," Patsy answered. "Look at this." She handed him a story she had torn out of a paper. A photograph with the

story showed a small statue. The caption under the picture read: "Helios, a small gold image of the Greek sun god was among relics stolen from the Cosmopolitan Museum. The figure of an Ibis, a sacred Egyptian bird, was also taken, along with some valuable old musical instruments."

"This man must be the thief," said Patsy. "He has stolen the treasures and is going to sell them to Bob Bell, Leo Hess, and Eli Hill, whoever they are. He wrote the deal out in code. The numbers on the right on his paper are the prices of the things. He wrote the rest of the deal in code, just in case he was suspected of anything crooked."

Patsy's mother picked up the telephone. "Patsy," she said, "I'm sure the police will be very interested."

So was the crook when they caught him.

TANK-THINK

"I have a remarkable water tank," said the farmer to the city man. "It weighs 27,867 pounds and it holds 25,831 gallons of water. Although it is full to the brim, I can still add three items that measure 1, 2, and 3 inches. What are the added items?

"I don't know," said the city fellow.

"You would if you had a calculator and could add all the numbers I gave you," said the farmer.

The city man didn't have a calculator, but you have. So what could the farmer add to his tank?

Answer: HOLES

You can make up reverse riddles to be solved by a calculator. But you don't start with the questions. You start with the answer—the word that appears in the display when you hold the calculator upside down. Then you make up the riddle and the calculations that are answered by the word.

The first step is to write the word backwards. Then you substitute digits for the letters. For example, if you want to get EELS in the display you write down SLEE. Look at the chart on page 53 that shows which numbers make letters, then write down the substitute digits—5,733.

Now work out a problem that adds up to that number. You could do it this way:

A fisherman's wife worked in the store where her husband sold his catch. She was 68 inches tall. Her arm span from the tip of the middle finger on her right hand to the tip of the middle finger on her left hand was also 68 inches. Her right arm was 24 inches from wrist to shoulder, and her left arm was 24 inches from wrist to shoulder. She measured 36 inches around the waist and 44 inches around the hips. Her neck was 16 inches around and her head was 20 inches around. Now, if you add all her measurements together, multiply by 20, and subtract 267 you can answer this question: what did she weigh?

Answer: 5,733.

After all, eels were something her husband's customers often bought from her.

So now you've played tricks and teased friends, and your little machine has one more upside-down message for you:

"Calculator fun is one of the best of 5318804"

INDEX

DATE			